Copyright

First edition: First printing
Illustrations and design © 2016
CptAim Gamer
All rights reserved. No part of this book may be reproduced or transmitted in any form or by any means, including but not limited to information storage and retrieval systems, electronic, mechanical, photocopy, recording, etc. without written permission from the copyright holder.
ISBN 978-1533385406

Lava Art

MATERIALS

OIL

GLASS

FOOD DYE
(ANY COLOR)

MATERIALS

SALT

WATER

STEP 1

POUR WATER!

Pour water into the glass

STEP 2

ADD FOOD DYE AND STIR!

Add a few drops of food dye

Stir it into the water

STEP 3

POUR THE OIL!

Pour your oil carefully in the water and you will see that it floats to the surface

Caution!!: Dont pour the oil too much

STEP 4

POUR THE SALT!

Take some salt and pour it into the water You will see that it sinks through the oil and landing heavily see what happen!

Tip: Try to drop a few drop of food dye on the oil and see what will happen!!

Caution: Put more salt if it stops creating bubbles

STEP 5

NOTE IT!

Note what did you see in this experiment

Lung Capacity

MATERIALS

 EMPTY LARGE BOTTLE

 FUNNEL

 JUG OF WATER

MATERIALS

 TAPE

 HOSE

 MARKER PEN

MATERIALS

 **BIG POT OF WATER**

 **MEASURING CUP**

STEP 1

TAKE YOUR TAPE!

Place your tape strip along the entire length of your large bottle

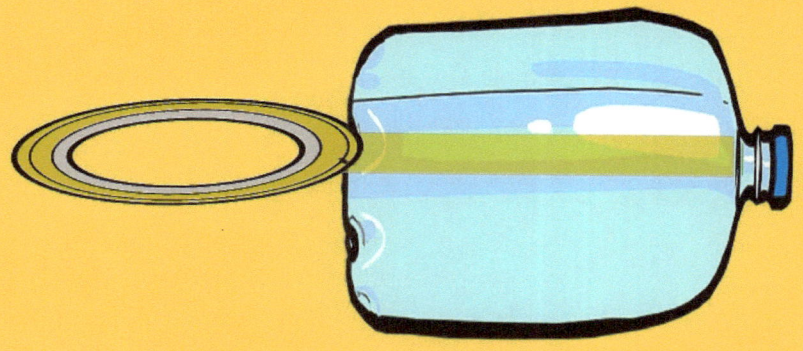

STEP 2

POUR THE WATER!

Take your funnel and put it in the neck of bottle

Pour 250 mm of water into the bottle

Take your marker pen and make a small line on the level of the water

You need to repeat this several times until the bottle full

STEP 3
FLIP IT!

Carefully turn it upside down in your pot

Put your hose in the bottle

Make sure that the water is not coming out when you do this step

STEP 4

BLOW IT!

Blow the air into the bottle as much as you can

STEP 5

USE THE FORMULA!

Count how many you got

$$(X*250)/1000 = \text{your lung capacity}$$

EXAMPLE

$$3*250 = 750 \text{mL then} /1000 = 0.75 \text{ litres}$$

WOW! This man got 0.75 litres of lung capacity!

Lava Lamp

MATERIALS

 **EMPTY BOTTLE**

 **FUNNEL**

 **VEGETABLE OIL**

MATERIALS

 WATER

 TORCH

 FOOD DYE
(any color)

MATERIALS

 ALKA SELTZER

 SMALL BOX
(WITH A HOLE)

STEP 1

POUR WATER!

Pour water into the bottle (not much), use your funnel to help!

STEP 3

PUT FOOD DYE IN THE BOTTLE!

Drop some food dye into the bottle

STEP 4

PUT SOME TABLET!

Drop some tablet of alka seltzer in the bottle

SEE WHAT HAPPEN!

STEP 5

LIGHT ON LIGHT OFF!

Put your opened torch inside small box, place the bottle on top of it, close the other light, and drop some alka seltza
SEE THE COOL THINGS!!

Water Defy Gravity

MATERIALS

 GALSS HALF FILLED WITH WATER

 PIECE OF CARD

 BUCKET

STEP 1

TAKE A CARD!

Take a card and put it on top of the glass firmly

Then swiftly turn the glass upside down

You wiil see a little bit of water escapes but otherwise everything stays inside

STEP 2

TAKE AWAY!

When you take away your hand you will see that the card is sucked onto the surface of the glass

Tip: This beacause gravity try to pull the water out of the glass but the card prevent the air from getting inside which cause a vacuum

STEP 3

LET AIR GO!

When you pull the card away you will see that the water go out insantly

Tip: This beacause air is able to go in side the glass and replace the water

Rainbow in a Tube

MATERIALS

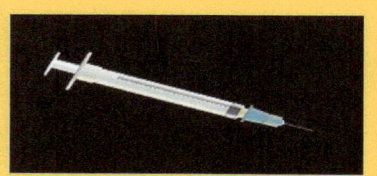 SYRINGE

 4 * GLASSES

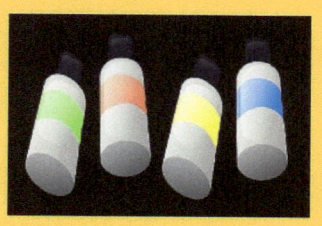 4 * FOOD DYE
(different color)

MATERIALS

TEST TUBE

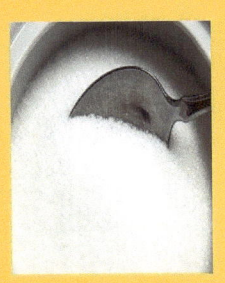

SUGAR

60 ML WATER X 4 (240ML)
(BOILED WATER IS BETTER)

STEP 1

ADD SUGAR!

In First glass add 1 tablespoon

In Second glass add 2 tablespoon

In Third glass add 3 tablespoon

In Fourth glass add 4 tablespoon

STEP 2

ADD WATER!

Add 60ml of water to each glass
(Boiling water reccomended)

STEP 3
STIR IT UP!

Stirring to make the sugar dissolve

STEP 4

DROP THE COLOR!

Glass no.1 add some Red food dye

Glass no.2 add some Yellow food dye

Glass no.3 add some Green food dye

Glass no.4 add some Blue food dye

STEP 5

FILL THE TEST TUBE !

First fill the blue color from forth glass
Second fill the green color from third glass
Third fill the Yellow from second glass
Fourth fill the Red from first glass

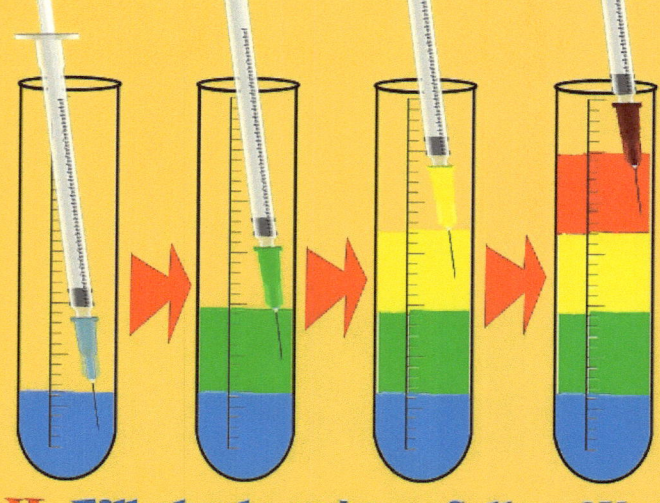

Caution!!: Fill slowly and carefully with syringe After fill a color finish, dont forget to clean the syringe

STEP 6

ENJOY YOUR SUCCESS!

Yeah!!! Congratulation!!!

What did you see when you hold the light up?!?

Tell everyone here

www.ingramcontent.com/pod-product-compliance
Lightning Source LLC
Chambersburg PA
CBHW042023200526
45159CB00035B/3033